Habitat Days **and Nights**

DAY AND NIGHT IN THE
Desert

by Ellen Labrecque

PEBBLE
a capstone imprint

Published by Pebble, an imprint of Capstone.
1710 Roe Crest Drive, North Mankato, Minnesota 56003
capstonepub.com

Library of Congress Cataloging-in-Publication Data
Names: Labrecque, Ellen, author.
Title: Day and night in the desert / by Ellen Labrecque.
Description: North Mankato, Minnesota : Pebble, [2022] | Series: Habitat days and nights | Includes bibliographical references and index. | Audience: Ages 5-8 | Audience: Grades K-1 | Summary: "Spend a day and night in the desert! Learn about this dry habitat through the unique animals that call it home. Stand with meerkats during a watchful hunt for insects. Spot a jackrabbit taking a shaded midday nap. Join a desert tortoise as it dines on colorful cactus fruit. After dark, follow a snake as it slithers in cool, moonlit sand. What will tomorrow bring in the desert?"-- Provided by publisher.
Identifiers: LCCN 2021041455 (print) | LCCN 2021041456 (ebook) |
  ISBN 9781663976949 (hardcover) | ISBN 9781666327632 (paperback) |
  ISBN 9781666327649 (pdf) | ISBN 9781666327663 (kindle edition)
Subjects: LCSH: Desert animals--Behavior--Juvenile literature. | Habitat (Ecology)--Juvenile literature.
Classification: LCC QL116 .L33 2022 (print) | LCC QL116 (ebook) | DDC 591.754--dc23
LC record available at https://lccn.loc.gov/2021041455
LC ebook record available at https://lccn.loc.gov/2021041456

Table of Contents

Words in **bold** are in the glossary.

What Is a Desert?

Deserts are dry **habitats**. They get less than 10 inches (25 centimeters) of rain per year. Most deserts are hot during the day. They are cool at night.

Many animals live in deserts. Some come out during the day. Others come out at night. There are deserts all over the world! One is the Sonoran Desert in North America.

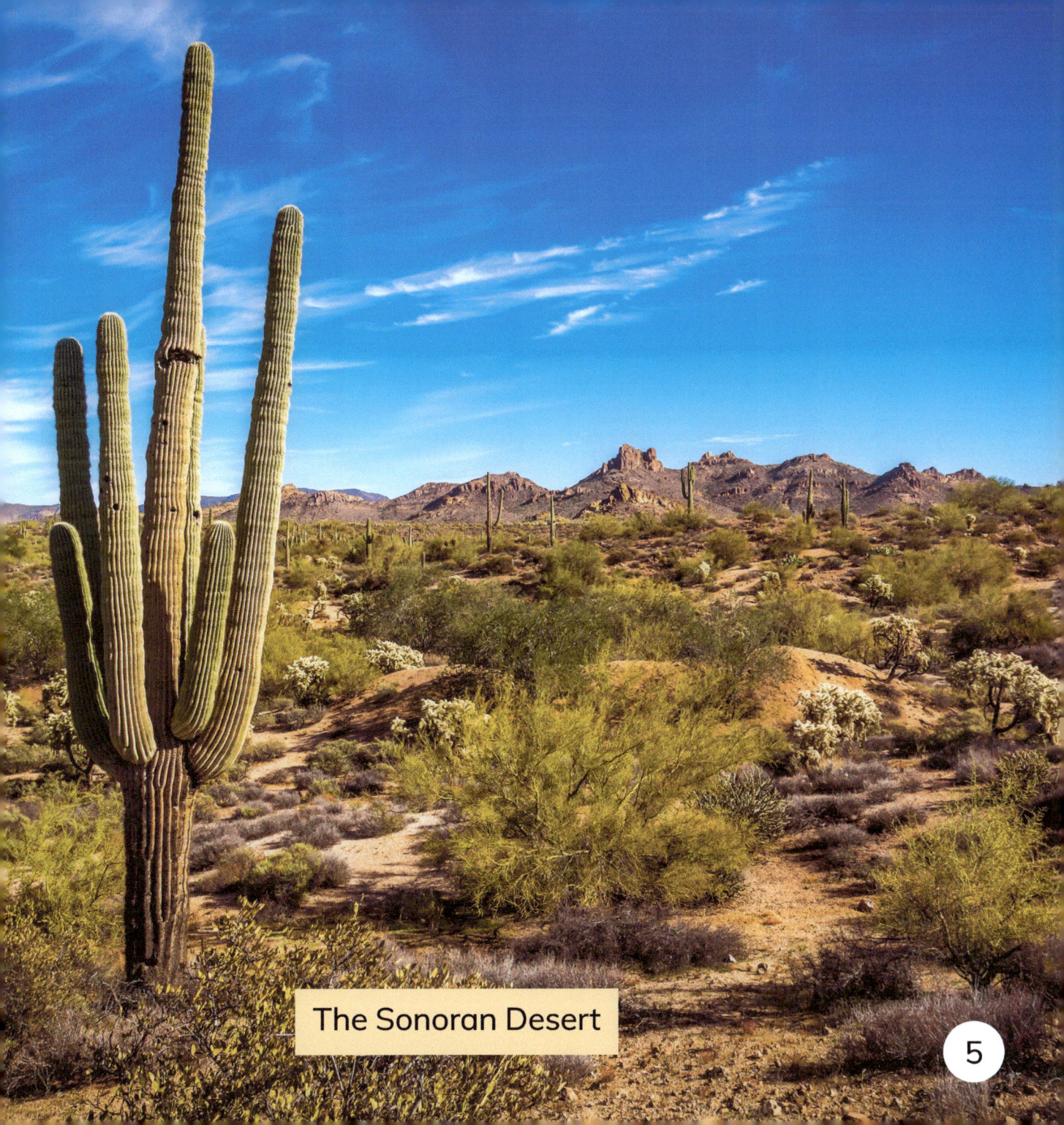

The Sonoran Desert

Morning

The sun rises in the Sonoran Desert. A desert tortoise comes out of its **burrow**. It will eat and drink while the air is cool. Then it will return to its burrow.

The tortoise walks on big, **scaly** legs. It drinks from a hole in the sand. The tortoise dug this hole to catch rainwater!

Desert tortoise

Noon

Animals try to stay cool in the heat. A jackrabbit finds shade under a bush. It will look for food at night. The jackrabbit eats grasses and **cacti**.

Jackrabbit

Cactus wren

A cactus wren hops under a large cactus. It turns over leaves to look for bugs. The brown bird moves slower as the air warms. It rests in the cool shade.

Late Afternoon

The sun warms the air and sand. An antelope squirrel shades itself with its tail. The squirrel climbs a cactus to eat cactus fruit. It avoids the **prickly** parts as it eats.

Antelope squirrel

Horned lizard

A horned lizard spent the morning eating ants. It caught the bugs with its sticky tongue. Now, the lizard rests in the shade.

Evening

A desert cottontail rabbit spent the afternoon under a bush. Now, the air is cool. The rabbit comes out. It eats grass and cacti with its sharp teeth.

Desert cottontail rabbit

Mule deer

A mule deer and its baby eat grass. They hear a sound. The deer run. They make short jumps on all four legs. This helps them move over rough ground.

Night

More animals wake as the air cools. A skunk comes out of its rocky burrow. It can't see well in the dark. But it has good senses of hearing and smell.

The skunk pushes its nose into the ground. It sniffs for food. A skunk will eat almost anything. This includes bugs, seeds, and eggs.

Skunk

Late Night

A tarantula searches for bugs to eat. A beetle crawls by. The tarantula grabs it with its long legs. It bites the beetle with its fangs.

Nearby, a night snake hunts for a meal. The snake grabs a frog in its jaws. The snake's **venom** knocks the frog out. Then, the snake swallows the frog whole!

Night snake

Tarantula

Dawn

A black-throated sparrow chirps from a cactus. Sparrows mostly spend time on the ground. But this morning, the males sit up high. They chirp to find **mates**.

Black-throated sparrow

Porcupine

A porcupine has spent the night eating leaves. It returns to its cool burrow. The porcupine rests as the sun rises. Another day in the desert has begun.

Lizard Activity

What You Need:

- popsicle stick
- orange marker
- 3 orange pipe cleaners
- scissors
- glue
- black marker

What You Do:

1. Color one side of a popsicle stick orange.

2. Cut a pipe cleaner in half. Glue both halves underneath the popsicle stick to make legs.

3. Cut another pipe cleaner in half. Cut the halves in half again. Twist the pipe cleaner pieces onto the ends of the legs to make toes.

4. Glue a third pipe cleaner to one end of the stick. This is the lizard's tail. Curl the end of the tail.

5. Draw two eyes on the other end of the stick in black marker.

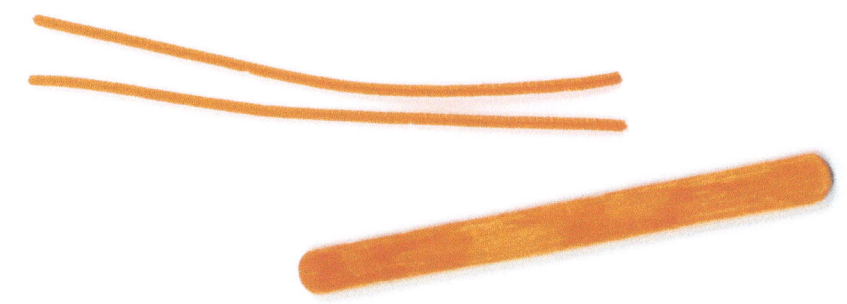

Glossary

burrow (BUHR-oh)—a tunnel or hole in the ground made or used by an animal

cactus (KACK-tuss)—a plant covered in spines that is found in desert areas

habitat (HAB-uh-tat)—the natural place and conditions in which a plant or animal lives

mate (MATE)—the male or female partner of an animal

prickly (PRIK-lee)—covered in small, sharp points

scaly (SKAY-lee)—covered in many small, hard pieces of skin called scales

venom (VEN-uhm)—poisonous liquid used by some snakes when they bite

Read More

Griffin, Annabel and Rose Maclachlan. *What Can I See in the Desert?* London: Hungry Tomato, 2021.

Keppeler, Jill. *20 Fun Facts about Desert Habitats*. New York: Gareth Stevens Publishing, 2022.

Topacio, Francine. *Creatures in a Hot Desert*. New York: PowerKids Press, 2020.

Internet Sites

Britannica Kids—Desert
kids.britannica.com/kids/article/desert/346108

DK Find Out!—Deserts
dkfindout.com/us/earth/deserts/

National Geographic Kids—Desert Habitat
kids.nationalgeographic.com/nature/habitats/article/desert

Index

About the Author

Ellen Labrecque is the author of more than 100 nonfiction children's books. She lives in Bucks County, Pennsylvania, with her husband and two kids. She has the best writing partner in the world—her dog, Oscar. An avid reader and runner, Ellen is a morning person. On most days, she is up before the sun.